何美好 編著

新手入廚系列

# 快、易滾湯

# 前言

廣東人愛飲湯來養生，不過人們總會把日常湯水與滋補湯或老火湯聯想在一起，彷似湯水只有這些。日常湯水中，蔬菜瓜果或海鮮作主要煮湯材料，也是不錯的選擇，因為它們可以在短時間成為省時美味的靚湯，既是清爽湯水，也可為佐餐美食，一舉數得，方便又簡單。

現今營養師也建議城市人飲食習慣以簡單輕便為主，所以滾湯或快煮湯是很好的選擇。特別是烹調時間應不超過 1 小時最為理想，因為利用短時間烹調食材，除了能保持營養價值，味道還不錯，又符合節省能源的環保概念。

許多新鮮材料都能在濕貨市場或超級市場內找到，一些時令蔬菜和海鮮，價錢便宜又品質優良，適合上班一族或雙職婦女作為煮菜靈感。書中輯錄了 40 多款簡捷滾湯作為主婦們或小家庭參考，相信必有一款適合大家的口味。

# 目錄



## 湯羹



## 海鮮類

## 蔬果類



## 家禽與肉類

4~6 人
Serves 4~6

25 分鐘
25 minutes

# 蟹肉魚翅瓜羹

# Crab Meat and Spaghetti Squash Thick Soup

## 材料 | Ingredients

蟹 1 隻
魚翅瓜 1 個
芫荽 1 棵
草菇（已氽水）40 克
雞蛋白 1 隻
清水 1 公升

1 crab
1 spaghetti squash
1 stalk coriander
40g straw mushrooms (scalded)
1 egg white
1 litre water

### 芡汁 | Thickening

生粉 1 茶匙
清水 4 湯匙

1 tsp cornstarch
4 tbsps water

### 調味料 | Seasonings

鹽少許
胡椒粉少許

pinch of salt
pinch of pepper

### 入廚貼士 | Cooking Tips

- 蟹可預先蒸熟，拆肉，放在冰箱中隨時可以應用。
- You may steam crab flesh in advance and keep in the refrigerator so that you may use when needed.

### 做法 | Method

1. 蟹劏洗乾淨，放蒸籠以大火蒸 10 分鐘，取出涼凍，拆肉。
2. 魚翅瓜去皮，用叉子把瓜肉拆出。備用。
3. 清水、草菇與魚翅瓜同置煲中煮 10 分鐘，放入蟹肉煮 1~2 分鐘。
4. 加入生粉水勾芡，熄火。加入雞蛋白拌勻，再加芫荽和調味便可。

1. Rinse crab thoroughly, steam in a steamer over high heat for 10 minutes, take out to cool, remove the flesh.
2. Peel spaghetti squash, remove the flesh with a fork, set aside.
3. Put straw mushrooms and spaghetti squash into a pot and add water, boil for 10 minutes. Add crab meat and cook for 1~2 minutes.
4. Add cornstarch solution to thicken the soup. Turn off the heat, add egg white and stir well. Add coriander and seasonings, serve.

# 雜菌素翅羹

# Imitation Shark's Fin Thick Soup with Assorted Mushrooms

## 材料 | Ingredients

鮮雜菌 150 克
素翅 100 克
腐竹 1/2 張
芥蘭梗丁粒 2 湯匙

150g assorted fresh mushrooms
100g imitation shark's fin
1/2 dried beancurd sheet
2 tbsps kale stem (diced)

4~6 人
Serves 4~6

45 分鐘
45 minutes

## 素湯底 | Ingredients of the Vegetarian Broth (A)

| | |
|---|---|
| 甘筍 80 克 | 80g carrot |
| 乾冬菇 40 克 | 40g dried black mushrooms |
| 粟米 1 條 | 1 stalk corn |
| 清水 1 1/2 公升 | 1 1/2 litres water |

## 芡汁 | Thickening

生粉 2 茶匙
清水 4 湯匙

2 tsps cornstarch
4 tbsps water

- 可改用罐裝清雞湯代替素湯底。
- You may use canned chicken broth to replace the veggie broth.

## 調味料 | Seasonings

鹽少許
胡椒粉少許

pinch of salt
pinch of pepper

## 做法 | Method

1. 素湯底材料洗淨，切塊，注入清水煲 30 分鐘，隔渣備用。
2. 鮮雜菌切粒，汆水，過冷。
3. 腐竹浸軟，瀝乾，與素湯同煮滾，加入鮮雜菌、素翅和芥蘭梗丁粒煮 3~5 分鐘。
4. 倒入生粉水勾芡，下調味，即成。

1. Rinse ingredients (A) thoroughly and cut into wedges. Put into a pot and add water, boil for 30 minutes, strain off the residue, set aside.
2. Dice assorted fresh mushrooms, scald and rinse with cold water.
3. Soak dried beancurd sheet till soft and drain. Cook dried beancurd sheet with the broth, add assorted fresh mushrooms, imitation shark's fin and diced kale stem, boil for 3~5 minutes.
4. Thicken the soup with cornstarch solution. Add seasonings, serve.

4~6 人
Serves 4~6

15 分鐘
15 minutes

# 粟米魚肚羹
# Corn and Fish Maw Thick Soup

### 材料 | Ingredients

| | |
|---|---|
| 粟米忌廉湯 1 罐 | 1 can corn cream soup |
| 已浸發魚肚 100 克 | 100g soaked fish maw |
| 雞蛋 1 隻 | 1 egg |
| 清水 500 毫升 | 500 ml water |

### 煨魚肚料 | Ingredients for Braising Fish Maw (A)

| | |
|---|---|
| 薑 2~3 片 | 2~3 slices ginger |
| 葱 2 條 | 2 stalks spring onion |
| 紹興酒 1 湯匙 | 1 tbsp Shaoxing wine |

## 調味料 | Seasonings

鹽 1/2 茶匙
胡椒粉適量

1/2 tsp salt
pinch of pepper (optional)

### 入廚貼士 | Cooking Tips

- 魚肚宜選完全爆透，以及不要揀顏色太白的，因為它可能被漂染，味道會淡一點。
- It would be better to choose thoroughly-fried fish maw. Don't choose those which are white in color as they may be bleached and the taste will be milder.

## 做法 | Method

1. 魚肚放入鍋中，加入適量水和煨魚肚料同煮 5 分鐘。
2. 取出魚肚，過冷，瀝乾水分，切丁粒。
3. 粟米忌廉湯加入清水同煮滾，加入魚肚丁粒煮 5~8 分鐘，熄火。
4. 雞蛋打勻，倒入粟米湯中拌勻，加入調味料便可。

1. Put fish maw into a pot, add some water and boil with the ingredients (A) for 5 minutes.
2. Take out fish maw, rinse with cold water, drain and dice.
3. Boil corn cream soup with water, add fish maw dice and boil for 5~8 minutes, remove from heat.
4. Whisk egg and add to the corn soup, stir well, add seasonings and serve.

# 韭黃瑤柱海參羹

# Dried Scallop and Sea Cucumber Soup with Yellow Chives

## 材料 | Ingredients

瑤柱碎 40 克
海參（已浸發）200 克
韭黃（切段）20 克
銀芽 20 克
甘筍絲 20 克
清水 1 公升

40g dried scallops
200g sea cucumber (soaked)
20g yellow chives (sectioned)
20g bean sprouts
20g shredded carrot
1 litre water

4~6 人
Serves 4~6

20 分鐘
20 minutes

## 煨海參料 | Ingredients for Braising Sea Cucumber (A)

| | |
|---|---|
| 薑 2~3 片 | 2~3 slices ginger |
| 葱 2 條 | 2 stalks spring onion |
| 紹興酒 1 湯匙 | 1 tbsp Shaoxing wine |

## 芡汁 | Thickening

馬蹄粉 4 湯匙
清水 4 湯匙

4 tbsps water chestnut powder
4 tbsps water

### 入廚貼士 | Cooking Tips

- 用馬蹄粉勾芡，口感會比較清爽。
- Use water chestnut powder as the thickening agent will give a fresher texture.

## 調味料 | Seasonings

鹽 1/2 茶匙
胡椒粉適量

1/2 tsp salt

pinch of pepper

## 做法 | Method

1. 瑤柱加 1/2 公升清水浸泡一夜，取出，撕碎，瑤柱水保留。
2. 海參加入適量水和煨海參料同煮 5 分鐘，取出，過冷，切絲。
3. 浸瑤柱水和清水同置鍋中燒滾，加入甘筍絲、瑤柱和海參煮 10 分鐘。
4. 倒入芡汁料推至濃稠，加入調味拌勻即可。

1. Soak dried scallops in water overnight, drain and tear into shreds. Reserve the soaking water.
2. Put sea cucumber into a pot and add water, boil with ingredients (A) for 5 minutes, take out, rinse with cold water, shred.
3. Put soaking water and water into a pot and bring to a boil, add shredded carrot, dried scallops and sea cucumber, cook for 10 minutes.
4. Thicken the soup with water chestnut solution. Add seasonings, stir well and serve.

4~6 人
Serves 4~6

20 分鐘
20 minutes

# 雞蓉冬蓉羹

## Chicken and Winter Melon Soup

### 材料 | Ingredients

| | |
|---|---|
| 冬瓜 300 克 | 300g winter melon |
| 雞胸肉 80 克 | 80g chicken brisket |
| 雞蛋白 1 隻 | 1 egg white |
| 清雞湯 500 毫升 | 500 ml chicken broth |
| 清水 400 毫升 | 400 ml water |

### 醃料 | Marinade

| | |
|---|---|
| 鹽 1/4 茶匙 | 1/4 tsp salt |
| 糖 1/2 茶匙 | 1/2 tsp sugar |
| 生粉 1/2 茶匙 | 1/2 tsp cornstarch |
| 油 1/2 茶匙 | 1/2 tsp oil |
| 清水 1 茶匙 | 1 tsp water |

## 芡汁 | Thickening

生粉 1 茶匙
清水 4 湯匙

1 tsp cornstarch
4 tbsps water

## 調味料 | Seasonings

鹽少許
胡椒粉少許

pinch of salt
pinch of pepper

### 入廚貼士 | Cooking Tips

- 熄火後才倒入蛋白，可避免蛋白變得粗糙，不夠滑嫩。
- Add egg white after removing from heat can make the egg white smooth.

## 做法 | Method

1. 雞胸肉剁碎，加入醃料拌勻。
2. 冬瓜洗淨，磨蓉。
3. 清雞湯與清水同置鍋中煮滾，加入冬瓜蓉煮 5 分鐘。
4. 放入芡汁推至濃稠，加入免治雞肉煮滾 2~3 分鐘或至雞肉熟透，熄火。
5. 雞蛋白拂勻，輕輕加入湯中，拌勻。下調味，便可享用。

1. Mince chicken brisket, add marinade and stir well.
2. Rinse winter melon thoroughly, grind to mash.
3. Put chicken broth and water into a pot, bring to a boil, add winter melon mash and cook for 5 minutes.
4. Add thickening and stir well. Add minced chicken and bring to a boil, cook for 2~3 minutes or until the chicken is well done. Remove from heat.
5. Whisk egg white, stir gently into the soup. Add seasonings and serve.

# 豆腐牛肉羹

## Beancurd and Beef Thick Soup

### 材料 | Ingredients

玉子豆腐（切粒）1 條
嫩豆腐（切粒）1 磚
免治牛肉 150 克
青豆（氽水）3 湯匙
雞蛋 1 隻
清雞湯 500 毫升
清水 500 毫升

1 strip egg beancurd (diced)
1 cube soft beancurd (dice)
150g minced beef
3 tbsps green pea (scalded)
1 egg
500 ml chicken broth
500 ml water

4~6 人
Serves 4~6

20 分鐘
20 minutes

湯羹
Thick Soup

### 醃料 | Marinade

| | |
|---|---|
| 薑汁 1 湯匙 | 1 tbsp ginger juice |
| 紹興酒 1 茶匙 | 1 tsp Shaoxing wine |
| 生粉 1 茶匙 | 1 tsp cornstarch |
| 油 1 茶匙 | 1 tsp oil |
| 醬油 1/2 茶匙 | 1/2 tsp soy sauce |
| 糖 1/2 茶匙 | 1/2 tsp sugar |

### 芡汁 | Thickening

馬蹄粉 4 湯匙
清水 4 湯匙

4 tbsps water chestnut powder
4 tbsps water

### 調味料 | Seasonings

鹽 1/2 茶匙
胡椒粉適量

1/2 tsp salt
pinch of pepper

### 做法 | Method

1. 免治牛肉加入醃料拌勻，備用。
2. 燒滾水一鍋，熄火，放入玉子豆腐粒和嫩豆腐粒浸熱，取出瀝乾。
3. 清雞湯和清水煮滾，把所有材料放進湯內煮 5 分鐘。
4. 加入芡汁煮至濃稠，下調味，熄火，加入雞蛋液拌勻。

1. Mix minced beef with marinade, stir well, set aside.
2. Heat a pot of water and remove from heat, add egg beancurd dice and soft bean curd dice, soak till hot and drain.
3. Bring chicken broth and water to a boil, add all ingredients and cook for 5 minutes.
4. Add sauce and cook till thickens, add seasonings. Remove from heat, add egg and stir well.

4~6 人
Serves 4~6

30 分鐘
30 minutes

# 烤鴨肉羹

# Roasted Duck Meat Thick Soup

## 材料 | Ingredients

| | |
|---|---|
| 燒鴨 1/2 隻 | 1/2 roasted duck |
| 蝦肉（切粒）40 克 | 40g shrimp meat (diced) |
| 中國芹菜（切絲）40 克 | 40g Chinese celery (shredded |
| 瑤柱（已蒸熟）20 克 | 20g dried scallops (steamed |
| 韭黃 20 克 | 20g yellow chives |
| 銀芽 20 克 | 20g bean sprouts |
| 金菇 40 克 | 40g enokitake mushrooms |
| 金華火腿絲適量 | some Jinhua ham shreds |
| 清水 1 1/2 公升 | 1 1/2 litres water |

### 芡汁 | Thickening

馬蹄粉 4 湯匙
清水 4 湯匙

4 tbsps water chestnut starch
4 tbsps water

### 調味料 | Seasonings

鹽 1/2 茶匙
胡椒粉適量

1/2 tsp salt
pinch of pepper

### 入廚貼士 | Cooking Tips

- 瑤柱浸水一夜，加入薑汁酒連浸瑤柱水同蒸，原水可作上湯使用。
- Soak dried scallops overnight, add ginger wine and steam together with the soaking water, the liquid can be used as broth.

### 做法 | Method

1. 燒鴨去皮，拆肉，鴨骨汆水，過冷。
2. 清水與燒鴨骨同置煲中煮 20 分鐘，取出鴨骨。
3. 放入鴨肉、瑤柱、金菇、蝦肉和中國芹菜絲煮滾，倒入芡汁推至濃稠。
4. 下調味拌勻，加入韭黃、銀芽和金華火腿拌勻，即可。

1. Skin and bone roasted duck. Scald duck bone and rinse with cold water. Reserve duck flesh.
2. Put duck bone and water in a pot, boil for 20 minutes. Discard duck bone.
3. Add duck meat, dried scallops, enokitake mushrooms, shrimp and shredded Chinese celery, bring to a boil. Thicken the soup with water chestnut starch solution.
4. Add seasoning and stir well. Add yellow chives, bean sprouts and Jinhua ham, stir well and serve.

4~6 人
Serves 4~6

25~30 分鐘
25~30 minutes

# 魚腐鮮蝦湯

# Fish Patties and Fresh Prawn Soup

## 材料 | Ingredients

大蝦 300 克
蝦膠 200 克
魚腐 10~12 個
甘筍（切絲）150 克
中國芹菜（切段）150 克
芫荽 2~3 棵
洋葱（切絲，爆香）1 個
清水 1 公升

300g king prawns
200g prawn paste
10~12 pcs fish patty
150g carrot (shredded)
150g Chinese celery (sectioned)
2~3 stalks coriander
1 onion (shredded, fried to fragrant)
1 litre water

### 醃料 | Marinade

| | |
|---|---|
| 糖 1 茶匙 | 1 tsp sugar |
| 生粉 1 茶匙 | 1 tsp cornstarch |
| 胡椒粉 1 茶匙 | 1 tsp pepper |
| 鹽 1/2 茶匙 | 1/2 tsp salt |
| 雞蛋白 1 隻 | 1 egg white |

### 做法 | Method

1. 大蝦洗淨，切頭、去殼，留尾。
2. 蝦膠加醃料拌勻，唧成蝦丸。
3. 熱鑊下油，放入蝦頭和蝦殼炒香，注入清水煮 10 分鐘，取出蝦頭和蝦殼，蝦湯留用。
4. 蝦湯加入甘筍絲、芫荽、洋葱和芹菜段煮 10 分鐘，加入魚腐、蝦丸和大蝦煮 3~5 分鐘便可。

1. Rinse prawns thoroughly, remove heads and shells and keep tails.
2. Add marinade to prawn paste, stir well, mould into balls.
3. Add oil to a heated wok, add prawn heads and shells, stir-fry till fragrant. Add water and cook for 10 minutes. Discard prawn heads and shells and retain prawn soup.
4. Add shredded carrot, coriander, onion and celery sections into prawn soup and cook for 10 minutes. Add fish patties, prawn balls and prawns and cook for 3~5 minutes, serve.

# 木瓜魚尾魚頭湯

## Fish Tail and Fish Head Soup with Papaya

### 材料 | Ingredients

木瓜 1 個
魚頭 1 個
魚尾 1 條
薑 1~2 片
清水 2 公升

1 papaya
1 fish head
1 fish tail
1~2 slices ginger
2 litres water

4~6 人
Serves 4~6

25~30 分鐘
25~30 minutes

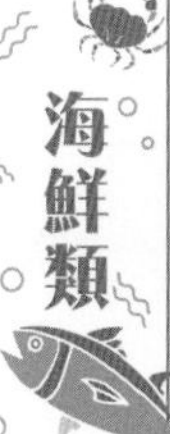


### 調味料 | Seasonings

鹽 1/2 茶匙
胡椒粉適量

1/2 tsp salt
pinch of pepper

### 入廚貼士 | Cooking Tips

- 揀選紅肉木瓜作此湯，味道更鮮甜。也可以加入鹹蛋增加食味。
- The taste will be more fresh and sweet if papaya with red flesh is used. You may add a salted egg to enhance the taste.

### 做法 | Method

1. 木瓜去皮，去籽，切塊。
2. 魚頭和魚尾洗淨，加 1 茶匙鹽擦勻全身。
3. 熱鑊下油，加入薑片炒香，放入魚頭和魚尾煎香，注入清水滾至呈奶白色。
4. 放入木瓜以大火煮 20~25 分鐘，下調味即可。

1. Rinse, peel and seed papaya, chop into pieces.
2. Rinse fish head and tail thoroughly, rub with 1 tsp of salt well.
3. Add oil to a heated wok, add ginger slices and stir-fry till fragrant. Add fish head and tail and sauté till fragrant, add water and cook until milky in color.
4. Add papaya and cook over high heat for 20~25 minutes, add seasonings and serve.

4~6 人
Serves 4~6

25~30 分鐘
25~30 minutes

# 勝瓜雜錦海鮮湯
# Assorted Seafood Soup with Angled Luffa

## 材料 | Ingredients

| | |
|---|---|
| 沙鯭魚 / 泥鯭魚 900 克 | 900g thread-sail file fish/ white spotted rabbitfish |
| 蜆 150 克 | 150g clams |
| 魷魚 1 條 | 1 squid |
| 蟹柳 4~5 條 | 4~5 crab sticks |
| 勝瓜 1 條 | 1 angled luffa |
| 草菇 150 克 | 150g straw mushrooms |
| 豆腐泡 6~8 個 | 6~8 beancurd puffs |
| 薑 40 克 | 40g ginger |
| 葱 2~3 條 | 2~3 stalks spring onion |
| 清水 1 1/2 公升 | 1 1/2 litres water |

## 調味料 | Seasonings

鹽 1/2 茶匙
胡椒粉適量

1/2 tsp salt
pinch of pepper

### 入廚貼士 | Cooking Tips

- 魚湯滾至大滾的作用，是可保持所有材料的色澤和鮮味。
- Bringing the fish soup to the boil is essential because this can keep the hue and fresh taste of all ingredients.

## 做法 | Method

1. 勝瓜去皮，切角。鮮菇汆水，過冷。
2. 沙鯭魚或泥鯭魚洗淨，起骨和起肉。
3. 熱鑊下油，放入薑葱爆香，加入魚骨煎香，灒酒，注入清水煮 20 分鐘，隔去魚骨。
4. 把原魚湯煮至大滾，加入所有材料煮 5~10 分鐘，下調味料，即成。

1. Peel angled luffa, chop into wedges. Scald straw mushrooms, rinse with cold water.
2. Rinse thread-sail file fish/ white spotted rabbitfish thoroughly, separate the bone and flesh.
3. Add oil to a heated wok, put in ginger and spring onion and stir-fry till fragrant. Add fish bone and sauté till fragrant. Sprinkle with wine, add water and cook for 20 minutes. Discard fish bone.
4. Bring fish soup to a boil, add all ingredients and cook for 5~10 minutes. Add seasonings and serve.

# 鮮魚蜆湯

## Fresh Fish and Clams Soup

### 材料 | Ingredients

| | |
|---|---|
| 蜆 600 克 | 600g clams |
| 鮮魚 600 克 | 600g fresh fish |
| 時菜 300 克 | 300g vegetables |
| 鮮冬菇 80 克 | 80g fresh mushrooms |
| 胡椒粒 1 茶匙 | 1 tsp pepper grains |
| 蒜頭 1 粒 | 1 clove garlic |
| 薑 1~2 片 | 1~2 slices ginger |
| 清水 1 1/2 公升 | 1 1/2 litre water |

4~6 人
Serves 4~6

25~30 分鐘
25~30 minutes

### 調味料 | Seasonings

鹽 1/2 茶匙
胡椒粉適量

1/2 tsp salt
pinch of pepper

#### 入廚貼士 | Cooking Tips

- 如果想蜆湯濃味一點，可把蜆和鮮魚同煮，不過蜆肉的鮮味會與魚湯融合，肉質會變得有點粗糙。
- You may boil clams and fish at the same time for stronger taste, however, the fresh taste of the clams will merge with the fish soup and the texture of the clams will be a bit tough.

### 做法 | Method

1. 蜆放入鹽水中浸 30 分鐘，擦洗乾淨。
2. 熱鑊下油，加入蒜頭、薑和胡椒粒炒香，加入蜆快手炒，灒酒，盛起。
3. 鮮魚洗淨，加入 1 茶匙鹽擦勻。熱鑊下油，加 1~2 片薑煎香，加入清水和鮮冬菇，以大火煮至變奶白。
4. 加入時菜和蜆煮 5~8 分鐘或熟便可，下調味即成。

1. Soak clams in salt water for 30 minutes, rub and rinse thoroughly.
2. Add oil to a heated wok, add garlic, ginger and pepper grains, stir-fry till fragrant. Add clams and stir-fry quickly, sprinkle with wine, dish up and set aside.
3. Rinse fish thoroughly, rub well with 1 tsp of salt. Add oil to a heated wok, add 1~2 slices of ginger, shallow-fry the fish till fragrant. Add water and mushrooms, cook over high heat until the soup turns milky.
4. Add vegetables and clams and cook for 5~8 minutes or until done. Add seasonings and serve.

4~6 人
Serves 4~6

15~25 分鐘
15~25 minutes

# 咖喱蟹湯

# Curry Crab Soup

**入廚貼士 | Cooking Tips**

- 芫荽頭具有獨特香味，所以應全棵放湯，味道更鮮美。
- Coriander stalk has a unique scent, so a whole stalk should be added to achieve a fresher taste.

## 材料 | Ingredients

| | |
|---|---|
| 花蟹 2 隻 | 2 coral crabs |
| 番茄 1~2 個 | 1~2 tomatoes |
| 洋葱（去衣）1 個 | 1 onion (peeled) |
| 鮮冬菇 50 克 | 50g fresh mushrooms |
| 薯仔（削皮）1 個 | 1 potato (peeled) |
| 香茅（切片）1 條 | 1 stalk lemongrass (sliced) |
| 芫荽 1~2 棵 | 1~2 stalks coriander |
| 蒜頭 1~2 粒 | 1~2 cloves garlic |
| 油咖喱 1 湯匙 | 1 tbsp oil curry |
| 清雞湯 500 毫升 | 500 ml chicken broth |

## 做法 | Method

1. 蟹劏洗乾淨，斬件，備用。
2. 洋葱、番茄、薯仔分別切塊，分別熱鑊下油爆香，備用。
3. 鮮冬菇汆水過冷，備用。
4. 熱鑊下油，加入蒜頭爆香，再加入油咖喱爆香，倒入蟹件爆炒片刻，灒酒。加入其他配料和清雞湯煮 15 分鐘，便可。

1. Rinse crabs thoroughly, chop into pieces, set aside.
2. Chop onion, tomatoes, potato into pieces separately. Add oil to a heated wok, fry to fragrant separately.
3. Scald mushrooms and rinse with cold water, set aside.
4. Add oil to a heated wok, add garlic, stir-fry till fragrant. Then add oil curry and stir-fry till fragrant, add crab pieces and stir-fry for a while, sprinkle with wine. Add other ingredients and chicken broth and cook for 15 minutes, serve.

# 番茄薯仔魚湯

## Fish Soup with Tomato and Potato

### 材料 | Ingredients

| | |
|---|---|
| 石狗公 / 鮮魚 900 克 | 900g rockfish/fresh fish |
| 番茄 3~4 個 | 3~4 tomatoes |
| 薯仔 1 個 | 1 potato |
| 洋葱 1 個 | 1 onion |
| 甘筍 1 條 | 1 carrot |
| 粟米 1 條 | 1 stalk corn |
| 中國芹菜 1 條 | 1 stalk Chinese celery |
| 薑 1~2 片 | 1~2 slices ginger |
| 蒜頭 1~2 粒 | 1~2 pcs garlic |
| 香葉 1~2 片 | 1~2 pcs bay leaves |

4~6 人
Serves 4~6

25~30 分鐘
25~30 minutes

### 調味料 | Seasonings

鹽 1/2 茶匙
胡椒粉適量

1/2 tsp salt
pinch of pepper

**入廚貼士 | Cooking Tips**

- 石狗公味道清鮮，如果改用大眼雞（木棉魚），味道則會更濃烈。
- Rockfish (Sebastiscus) tastes fresh while the taste will be stronger if big eye fish (Priacanthus) is used instead.

### 做法 | Method

1. 所有蔬菜材料洗淨，切塊。
2. 熱鑊下油，加入蒜頭炒香，放入所有雜菜略炒片刻，加入香葉煮片刻，盛起。
3. 石狗公洗淨，加入 1 茶匙鹽擦勻。熱鑊下油，加 1~2 片薑和石狗公魚一同煎香，加入清水以大火煮至變奶白色。
4. 加入雜菜以大火煮 25~30 分鐘，即可。

1. Rinse all the vegetable ingredients thoroughly, chop into pieces.
2. Add oil to a heated wok, add garlic, stir-fry till fragrant. Add all vegetables, stir-fry roughly for a while, add bay leaves and cook for a while, dish up and set aside.
3. Rinse rockfish thoroughly, rub well with 1 tsp of salt. Add oil to a heated wok, add 1~2 slices of ginger and sauté with rockfish till fragrant. Add water and cook over high heat until the soup turns milky.
4. Add vegetables, cook over high heat for 25~30 minutes and serve.

4~6 人
Serves 4~6

25~30 分鐘
25~30 minutes

# 烏豆魚頭湯

## Fish Head Soup with Black Soy Beans

### 材料 | Ingredients

大魚頭 1 個
黑豆 80 克
瘦豬肉 60 克
薑絲 40 克
木耳（浸發）20 克
雞蛋 1 隻

1 big fish head
80g black soy beans
60g lean pork
40g shredded ginger
20g dried black wood ear fungus (soaked)
1 egg

## 醃料｜Marinade

| | |
|---|---|
| 薑汁 1 湯匙 | 1 tbsp ginger juice |
| 紹興酒 1 茶匙 | 1 tsp Shaoxing wine |
| 生粉 1 茶匙 | 1 tsp cornstarch |
| 油 1 茶匙 | 1 tsp oil |
| 鹽 1/2 茶匙 | 1/2 tsp salt |
| 糖 1/2 茶匙 | 1/2 tsp sugar |
| 清水 1 湯匙 | 1 tbsp water |

## 做法｜Method

1. 瘦豬肉切片，加入醃料拌勻。
2. 黑豆白鑊炒香，盛起。
3. 大魚頭洗淨，加鹽擦勻，起鑊下油，加入薑片，放入魚頭煎至金黃，撥放鑊旁。
4. 加點油爆香薑絲、木耳和瘦肉，打入雞蛋煎熟，灒酒，加入清水 1 公升煮 15~20 分鐘，即成。

1. Slice lean pork, add marinade and stir well.
2. Stir-fry black soy beans in a clean wok till fragrant, dish up and set aside.
3. Rinse big fish head thoroughly, add salt and rub well. Add oil to a heated wok, add ginger slices, put in the fish head and sauté till fragrant. Set aside in the wok.
4. Add some oil, stir-fry shredded ginger, black fungus and lean pork till fragrant. Add egg and sauté till done, sprinkle with wine. Add 1 litre of water and boil for 15~20 minutes, serve.

# 生菜鯪魚球湯

# Dace Balls Soup with Lettuce

## 材料 | Ingredients

鯪魚骨 450 克
免治鯪魚肉 300 克
冬菇（已浸泡）40 克
蝦米 20 克
生菜 1 棵
薑 2~3 片
葱 1~2 棵
清水 1 1/2 公升

450g dace bone
300g minced dace flesh
40g dried black mushrooms (soaked)
20g dried shrimps
1 stalk lettuce
2~3 slices ginger
1~2 stalks spring onion
1 1/2 litres water

4~6 人
Serves 4~6

30~40 分鐘
30~40 minutes

### 醃料 | Marinade

雞蛋白 1 隻
糖 1 茶匙
生粉 1 茶匙
胡椒粉 1 茶匙
鹽 1/2 茶匙
麻油少許
清水 1 湯匙

1 egg white
1 tsp sugar
1 tsp cornstarch
1 tsp pepper
1/2 tsp salt
some sesame oil
1 tbsp water

### 調味料 | Seasonings

鹽 1/2 茶匙
胡椒粉適量

1/2 tsp salt
pinch of pepper

### 做法 | Method

1. 生菜洗淨，切絲。
2. 免治鯪魚肉加入醃料拌勻，攪成有彈性和黏度。
3. 熱鑊下油，放入薑葱和蝦米爆香，加入魚骨煎香，灒酒。注入清水和冬菇煮 30 分鐘，隔去魚骨。
4. 放鯪魚球煮至浮起，下調味，加入生菜絲即成。

1. Rinse lettuce thoroughly, shred.
2. Add marinade to minced dace flesh and mix well, stir to springy and sticky.
3. Add oil to a heated wok, put in ginger, spring onion and dried shrimps, stir-fry till fragrant. Add fish bone and sauté till fragrant, sprinkle with wine. Add water and black mushrooms and cook for 30 minutes, discard fish bone.
4. Add dace balls and cook until floats. Add seasonings and shredded lettuce, serve.

4~6 人
Serves 4~6

25~30 分鐘
25~30 minutes

# 蝦米粉絲梭羅魚湯

# Glassy Perchlet Soup with Dried Shrimps and Green Bean Vermicelli

## 材料 | Ingredients

| | |
|---|---|
| 梭羅魚 600 克 | 600g glassy perchlet |
| 蝦米 40 克 | 40g dried shrimps |
| 甘筍 1 條 | 1 carrot |
| 洋葱 1 個 | 1 onion |
| 翠肉瓜 1 條 | 1 jade melon |
| 粉絲（已浸泡）80 克 | 80g green bean vermicelli (soaked) |
| 皮蛋（去殼切片）1~2 隻 | 1~2 preserved eggs (shell removed and sliced) |
| 薑 2~3 片 | 2~3 slices ginger |
| 葱花 1~2 條 | 1~2 stalks spring onion |

## 調味料 | Seasonings

| | |
|---|---|
| 鹽 1/2 茶匙 | 1/2 tsp salt |
| 胡椒粉適量 | pinch of pepper |

## 做法 | Method

1. 甘筍和翠肉瓜洗淨，切塊；洋葱洗淨，切塊。
2. 梭羅魚洗淨，加入 1 茶匙鹽擦勻。
3. 熱鑊下油，加 1~2 片薑、蝦米和梭羅魚一同煎香，放入魚袋，加入甘筍、洋葱和清水以大火煮至變奶白。
4. 加入翠肉瓜煮 5~10 分鐘，再加入粉絲和皮蛋煮滾，下調味即成。

1. Rinse carrot and jade melon thoroughly, chop into pieces. Rinse onion thoroughly and chop into pieces.
2. Rinse glassy perchlet thoroughly and rub well with 1 tsp of salt.
3. Add oil to a heated wok, add 1~2 slices of ginger, dried shrimps and glassy perchlet, sauté till fragrant together and put in a white cloth bag. Add carrot, onion and water, cook over high heat until the soup turns milky.
4. Add jade melon and cook for 5~10 minutes, add vermicelli and preserved eggs and bring to a boil. Add seasonings and serve.

# 時菜雲吞海鮮湯

# Wontons and Seafood Soup with Vegetables

## 材料 | Ingredients

| | |
|---|---|
| 鮮魚 600 克 | 600g fresh fish |
| 蜆 300 克 | 300g clams |
| 蝦 4~6 隻 | 4~6 shrimps |
| 蟹 1 隻 | 1 crab |
| 魷魚 1 隻 | 1 squid |
| 雲吞 12 隻 | 12 wontons (Cantonese shrimp dumpling) |
| 時菜 300 克 | 300g seasonal vegetables |
| 薑 1~2 片 | 1~2 slices ginger |
| 清水 2 公升 | 2 litres water |

4~6 人
Serves 4~6

25~30 分鐘
25~30 minutes

## 調味料 | Seasonings

鹽 1/2 茶匙
胡椒粉適量

1/2 tsp salt
pinch of pepper

## 做法 | Method

1. 魷魚洗淨，切塊，剞花。蟹洗淨，去除內臟，斬件。
2. 煮滾一鍋水，加入雲吞，不用蓋鑊蓋，煮至浮起，加入 1 杯凍水，待再滾起時，盛起備用。
3. 鮮魚洗淨，加入 1 茶匙鹽擦勻。熱鑊下油，下薑片和鮮魚一同煎香，加入清水以大火煮至變奶白。
4. 放入蟹煮 10 分鐘，再放入其餘海鮮和時菜煮 5~10 分鐘，最後放入雲吞煮滾，下調味便可。

1. Rinse squid thoroughly, chop into pieces and cut crisscross pattern. Rinse crab thoroughly, remove internal organs and chop into pieces.
2. Boil wontons in a pot of boiling water without lid until float. Add a cup of cold water and wait until it boils again, dish up and set aside.
3. Rinse fish thoroughly, rub well with 1 tsp of salt. Add oil to a heated wok, add ginger slices, sauté fish till fragrant. Add water and boil over high heat until the soup turns milky.
4. Add crab and cook for 10 minutes. Then add other seafood and vegetables and cook for 5~10 minutes. Lastly add wontons, bring to a boil, add seasonings and serve.

4~6 人
Serves 4~6

25~30 分鐘
25~30 minutes

# 四寶紫菜魚乾湯

## Fried Small Fish Soup with Four Kinds of Fish Balls and Dried Seaweed

### 材料 | Ingredients

鯪魚骨 300 克
小魚乾 150 克
四寶丸（任何四款）16 粒
紫菜 1 片
蝦乾 40 克
胡椒粒 1 茶匙
芫荽 1~2 個
薑 1~2 片
清水 2 公升

300g dace bone
150g fried small fishes
16 pcs any four kinds of fish balls
1 sheet dried seaweed
40g dried prawns
1 tsp pepper grains
1~2 stalks coriander
1~2 slices ginger
2 litres water

### 調味料 | Seasonings

冬菜適量
芹菜粒適量
炸香蒜片適量
葱粒適量
胡椒粉適量

some preserved Chinese cabbage
some celery dice
some deep-fried garlic slices
some spring onion dice
pinch pepper

### 做法 | Method

1. 鯪魚骨洗淨，加入 1/2 茶匙鹽擦勻。熱鑊下油，加薑片、胡椒粒和魚骨一同煎香，加入清水、魚乾和蝦乾以大火煮至變奶白。
2. 放入魚丸煮至浮起。
3. 把紫菜放碗中，倒入魚湯，吃時伴以調味料享用。

1. Rinse dace bone thoroughly, rub well with 1/2 tsp of salt. Add oil to a heated wok, add ginger slices, pepper grains and sauté with dace bone till fragrant. Add water, fried small fishes and dried prawns, cook over high heat until the soup turns milky.
2. Add fish balls and boil until float.
3. Put dried seaweed into a bowl and pour in fish soup. Serve with seasonings.

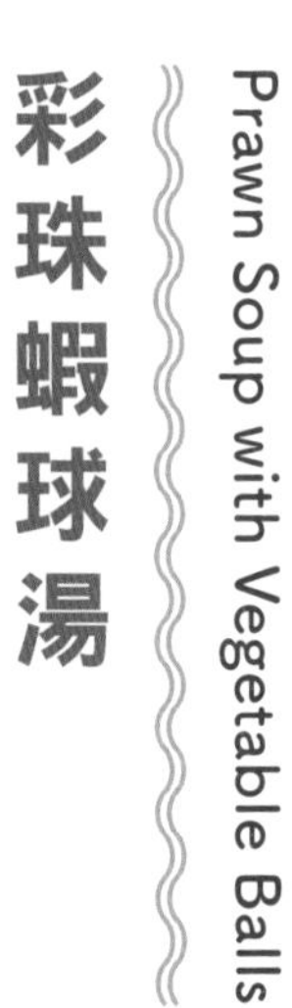

# 彩珠蝦球湯

## Prawn Soup with Vegetable Balls

4~6 人
Serves 4~6

25~30 分鐘
25~30 minutes

## 材料 | Ingredients

| | |
|---|---|
| 大蝦 6~12 隻 | 6~12 king prawns |
| 蝦子 1 湯匙 | 1 tbsp prawn roe |
| 合掌瓜 / 冬瓜 300 克 | 300g chayote (closed palms squash)/ winter melon |
| 青蘿蔔 300 克 | 300g green radish |
| 甘筍 1 條 | 1 carrot |
| 蒜頭 1~2 粒 | 1~2 cloves garlic |
| 芫荽 2~3 棵 | 2~3 stalks coriander |
| 胡椒粒 1 茶匙 | 1 tsp pepper grains |
| 清水 1 1/2 公升 | 1 1/2 litres water |

## 做法 | Method

1. 合掌瓜（冬瓜）、青蘿蔔和甘筍同置煲中煮軟，挖成球狀。
2. 大蝦洗淨，切頭、去殼，留尾。
3. 熱鑊下油，放入蒜頭、胡椒粒、蝦頭、蝦殼和蝦子爆香，注入清水以大火煮 15 分鐘。
4. 加入芫荽、合掌瓜、青蘿蔔和甘筍球煮 10 分鐘，放入大蝦煮熟即可。

1. Put chayote (winter melon), green radish and carrot into a pot, cook until soft, dig into balls.
2. Rinse prawns thoroughly, remove heads and shells, keep the tails.
3. Add oil to a heated wok, add garlic, pepper grains, prawn heads, shells and prawn roe, stir-fry until fragrant. Add water and cook over high heat for 15 minutes.
4. Add coriander, balls of chayote, green radish and carrot and cook for 10 minutes. Add prawns and cook until done, serve.

4~6 人
Serves 4~6

15 分鐘
15 minutes

# 節瓜鹹蛋肉片湯

# Salted Eggs and Pork Slice Soup with Hairy Melon

## 材料 | Ingredients

節瓜 600 克
瘦豬肉 200 克
鹹蛋 2 隻
薑 1 塊
清水 2 公升

600g hairy melon
200g lean pork
2 salted eggs
1 pc ginger
2 litres water

## 醃料 | Marinade

薑汁 1 湯匙
鹽 1/2 茶匙
糖 1/2 茶匙
紹興酒 1 茶匙
生粉 1 茶匙
油 1 茶匙

1 tbsp ginger juice
1/2 tsp salt
1/2 tsp sugar
1 tsp Shaoxing wine
1 tsp cornstarch
1 tsp oil

### 入廚貼士 | Cooking Tips

- 節瓜可用筷子刮皮，不應削皮，否則節瓜的味道會變得不夠嫩滑。
- You may scrape the skin of the hairy melon with a chopstick instead. However, the taste of the hairy melon will be not smooth enough if peels.

## 做法 | Method

1. 瘦豬肉洗淨，切片，加入醃料拌勻。
2. 節瓜用小刀刮皮，洗淨，切塊。
3. 節瓜、清水和薑同置煲中以大火煮 20 分鐘。
4. 加入瘦豬肉煮 10 分鐘，打入鹹蛋煮改中火煮 10 分鐘，熄火，焗 5 分鐘，即成。

1. Rinse lean pork thoroughly and slice. Add marinade and stir well.
2. Scrape hairy melon's skin with a small knife. Rinse thoroughly and cut into pieces.
3. Put hairy melon, water and ginger into a pot, cook over high heat for 20 minutes.
4. Add lean pork and cook for 10 minutes. Add salted egg, reduce to medium heat and cook for 10 minutes. Remove from heat and leave for 5 minutes and serve.

# 潺菜鹹魚頭豆腐湯

## Salted Fish Head and Beancurd Soup with Basella Alba (Ceylon Spinach)

### 材料 | Ingredients

鹹魚頭（小）1/2 個
魚尾 1 條（約 450 克）
潺菜 300 克
豆腐 1 磚
薑 1~2 片
滾水 1 1/2 公升

1/2 salted fish head (small)
1 fsh tail (about 450g)
300g basella alba (Ceylon spinach)
1 cube beancurd
1~2 slices ginger
1 1/2 litres boiling water

4~6 人
Serves 4~6

30~40 分鐘
30~40 minutes

### 調味料 | Seasonings

胡椒粉

pinch of pepper

**入廚貼士 | Cooking Tips**

- 鹹魚頭味道已很鹹，不用下鹽調味。
- 沒有潺菜，可改用其他時菜。
- Salted fish head is pretty salty already, hence, no need to add salt for seasoning.
- If basella alba is not available, other seasonal vegetables may be used to replace.

### 做法 | Method

1. 魚尾洗淨，加 1/2 茶匙鹽擦勻全身。熱鑊下油，下薑片煎香，倒入 250 毫升滾水煮至奶白。
2. 鹹魚頭洗淨，熱鑊下油煎香，轉放入魚尾處煮 5 分鐘，加入其餘滾水煮 20 分鐘。
3. 加入豆腐和潺菜煮至軟脸，需時約 5~10 分鐘。

1. Rinse fish tail thoroughly, rub well with 1/2 tsp of salt. Add oil to a heated wok, add ginger slices, shallow-fry until fragrant. Pour in 250 ml of boiling water and boil until the soup turns milky in color.
2. Rinse salted fish head thoroughly. Add oil to a hot wok and shallow-fry until fragrant. Add to fish tail soup and boil for 5 minutes. Add the remaining portion of boiling water and boil for 20 minutes.
3. Add beancurd and basella alba and boil for about 5~10 minutes until basella alba turns soft.

4~6 人
Serves 4~6

15 分鐘
15 minutes

# 南瓜粟米肉粒湯

## Pumpkin and Corn Soup with Pork Dice

### 材料 | Ingredients

南瓜 600 克
瘦豬肉 150 克
竹笙（已浸發）100 克
新鮮粟米粒 50 克
洋葱（切丁粒）1 個
清水 1 1/2 公升

600g pumpkin
150g lean pork
100g bamboo fungus (soaked)
50g fresh corn grains
1 onion (diced)
1 1/2 litres water

## 醃料 | Marinade

| | |
|---|---|
| 薑汁 1 湯匙 | 1 tbsp ginger juice |
| 紹興酒 1 茶匙 | 1 tsp Shaoxing wine |
| 生粉 1 茶匙 | 1 tsp cornstarch |
| 油 1 茶匙 | 1 tsp oil |
| 鹽 1/2 茶匙 | 1/2 tsp salt |
| 糖 1/2 茶匙 | 1/2 tsp sugar |

## 煨料 | Ingredients for Braising

| | |
|---|---|
| 薑汁 1 湯匙 | 1 tbsp ginger juice |
| 酒 1 茶匙 | 1 tsp wine |
| 葱 1 條 | 1 stalk spring onion |

## 做法 | Method

1. 南瓜去皮和去籽，切丁粒。
2. 瘦豬肉洗淨，切丁粒，加入醃料拌勻。
3. 竹笙洗淨，放入煨料煮 5 分鐘，取出飛水過冷，瀝乾，切粒。
4. 熱鑊下油，加入洋葱粒炒香，注入清水和放入南瓜粒同煮 20 分鐘。
5. 加入其他材料煮 5~10 分鐘，即成。

1. Peel pumpkin and remove seeds, dice.
2. Rinse lean pork thoroughly and dice, add marinade and stir well.
3. Rinse bamboo fungus thoroughly, add braising ingredients and boil for 5 minutes. Take out and scald. Rinse with cold water, drain and dice.
4. Add oil to a heated wok, add onion dice and stir-fry until fragrant. Add water and pumpkin dice and boil together for 20 minutes.
5. Add remaining ingredients and boil for 5~10 minutes, serve.

# 番茄肉碎蛋花湯

## Tomato and Minced Meat Soup with Egg

### 材料 | Ingredients

番茄 600 克
免治豬肉 150 克
薏米（已浸泡 30 分鐘）50 克
洋葱（切絲）1 個
芫荽 1 棵
蒜蓉 1 茶匙
雞蛋 1~2 隻
清水 1 1/2 公升

600g tomatoes
150g minced pork
50g job's tears (soaked for 30 mins)
1 onion (shredded)
1 stalk coriander
1 tsp minced garlic
1~2 eggs
1 1/2 litres water

4~6 人
Serves 4~6

25~30 分鐘
25~30 minutes

### 醃料 | Marinade

薑汁 1 湯匙
紹興酒 1 茶匙
生粉 1 茶匙
油 1 茶匙
鹽 1/2 茶匙
糖 1/2 茶匙
清水 1 湯匙

1 tbsp ginger juice
1 tsp Shaoxing wine
1 tsp cornstarch
1 tsp oil
1/2 tsp salt
1/2 tsp sugar
1 tbsp water

### 入廚貼士 | Cooking Tips

- 熄火才加入雞蛋液，拌勻時才容易造成蛋絲效果，可確保雞蛋滑嫩。
- Add egg liquid after removing from heat makes better result for forming egg shreds and ensure that the eggs being smooth enough.

### 做法 | Method

1. 免治豬肉加入醃料拌勻。
2. 番茄洗淨，用水沖去番茄籽，一開四份。
3. 熱鑊下油加入洋葱炒香，再加入番茄炒至軟身。注入清水和加入整棵芫荽，以大火煮 20 分鐘。
4. 取出芫荽，加入免治豬肉煮 5~10 分鐘，待免治豬肉全熟，熄火，打入雞蛋液，即可上桌。

1. Add marinade to minced pork and stir well.
2. Rinse tomatoes thoroughly, remove seeds by running water, cut into small pieces.
3. Add oil to a heated wok, add onion and stir-fry until fragrant. Add tomatoes and stir-fry until soft. Add water and the whole stalk of coriander and boil over high heat for 20 minutes.
4. Discard coriander, add minced pork and boil for 5~10 minutes until well done. Remove from heat, add egg liquid and stir well, serve.

4~6 人
Serves 4~6

25 分鐘
25 minutes

# 芥菜番薯窩蛋湯

# Mustard Greens, Sweet Potato and Egg Soup

### 材料 | Ingredients

芥菜 300 克
番薯 300 克
雞蛋 2 隻
薑 2~3 片
清雞湯 500 毫升
清水 500 毫升

300g mustard greens
300g sweet potatoes
2 eggs
2~3 slices ginger
500 ml chicken broth
500 ml water

### 調味料 | Seasonings

鹽 1/2 茶匙

1/2 tsp salt

**入廚貼士 | Cooking Tips**

- 不喜歡雞湯的話，可只用清水，但味道會略嫌單調。
- You may use water instead of chicken broth but the taste may be a bit too dull.

### 做法 | Method

1. 芥菜洗淨切段。番薯去皮，切塊。
2. 熱鑊下油 1 湯匙，放入薑片爆香，加入番薯炒片刻，注入清雞湯和清水煮 10~15 分鐘。
3. 加入芥菜再煮 10 分鐘，熄火。打入雞蛋浸熟，下調味便可。

1. Rinse mustard greens thoroughly and cut into sections. Peel sweet potatoes and cut into pieces.
2. Add 1 tbsp of oil to a heated wok, add ginger slices and stir-fry until fragrant. Add sweet potato and stir-fry for a while. Add chicken broth and water and boil for 10~15 minutes.
3. Add mustard greens and boil for another 10 minutes, remove from heat. Add eggs and leave until done in the hot soup. Add seasonings and serve.

# 南瓜素翅湯

# Pumpkin and Imitation Shark's Fin Soup

## 材料 | Ingredients

| | |
|---|---|
| 南瓜 600 克 | 600g pumpkin |
| 素翅 300 克 | 300 imitation shark's fin |
| 雞胸肉（切幼絲）150 克 | 150g chicken brisket (finely shredded) |
| 粟米粒 50 克 | 50g corn grains |
| 罐裝紅菜頭（切粒）50 克 | 50g can beetroot (diced) |
| 中國芹菜 50 克 | 50g Chinese celery |
| 蟹柳 2~3 條 | 2~3 imitation crab sticks |
| 清雞湯 500 毫升 | 500 ml chicken broth |
| 清水 400 毫升 | 400 ml water |

4~6 人
Serves 4~6

30 分鐘
30 minutes

### 醃料 | Marinade

雞蛋白 1 隻
生粉 1 茶匙
油 1 茶匙
鹽 1/2 茶匙
糖 1/2 茶匙

1 egg white
1 tsp cornstarch
1 tsp oil
1/2 tsp salt
1/2 tsp sugar

### 調味料 | Seasonings

鹽 1/2 茶匙

1/2 tsp salt

### 做法 | Method

1. 中國芹菜洗淨，切粒。
2. 蟹柳略沖水，撕條。雞肉加醃料拌勻備用。
3. 南瓜去皮和去籽，洗淨切塊，加入清水和清雞湯同煮至腍軟。
4. 加入雞肉絲和素翅煮滾。再倒入其他材料煮滾，下調味料便可。

1. Rinse Chinese celery thoroughly and dice.
2. Roughly rinse imitation crab sticks, tear into strips. Add marinade to chicken meat, stir well and set aside.
3. Peel pumpkin and remove seeds, rinse thoroughly, cut into pieces. Put pumpkin into a pot, add water and chicken broth, bring to a boil and cook until soft.
4. Add chicken meat shreds and imitation shark's fin, bring to a boil. Add remaining ingredients and bring to a boil. Add seasoning to taste.

4~6 人
Serves 4~6

15 分鐘
15 minutes

# 雜菌豆腐湯
## Beancurd Soup with Assorted Mushrooms

### 材料 | Ingredients

| | |
|---|---|
| 草菇 100 克 | 100g straw mushrooms |
| 冬菇 100 克 | 100g black mushrooms |
| 蟹菇 100 克 | 100g crab mushrooms |
| 金菇 100 克 | 100g enokitake mushrooms |
| 時菜 150 克 | 150g seasonal vegetables |
| 日本豆腐 1 盒 | 1 box Japanese beancurd |
| 瘦豬肉 150 克 | 150g lean pork |
| 清雞湯 500 毫升 | 500 ml chicken broth |
| 清水 500 毫升 | 500 ml water |

## 醃料 | Marinade

| | |
|---|---|
| 紹興酒 1 茶匙 | 1 tsp Shaoxing wine |
| 油 1 茶匙 | 1 tsp oil |
| 生粉 1 茶匙 | 1 tsp cornstarch |
| 醬油 1/2 茶匙 | 1/2 tsp soy sauce |
| 鹽 1/2 茶匙 | 1/2 tsp salt |
| 糖 1/2 茶匙 | 1/2 tsp sugar |

## 調味料 | Seasonings

鹽 1/2 茶匙

1/2 tsp salt

## 做法 | Method

1. 瘦豬肉洗淨，切片，加入醃料拌勻。
2. 草菇洗淨，汆水，過冷備用。
3. 草菇、清水和清雞湯同置煲中煮滾，放入肉片和其他材料（除豆腐外）煮滾。
4. 加入豆腐煮滾，熄火，下調味料便可。

1. Rinse lean pork thoroughly, slice, add marinade and stir well.
2. Rinse straw mushrooms thoroughly, scald, rinse with cold water, set aside.
3. Put straw mushrooms, water and chicken broth into a pot, bring to a boil. Add lean pork and the remaining ingredients except beancurd, bring to a boil again.
4. Add beancurd, bring to a boil, add salt to taste and serve.

# 枸杞豬肝湯

## Matrimony Vine and Pig's Liver Soup

### 材料 | Ingredients

枸杞 450 克
豬肝 200 克
薑 2~3 片
雞蛋 1~2 隻
清水 2 公升

450g matrimony vine
200g pig's liver
2~3 slices ginger
1~2 eggs
2 litres water

4~6 人
Serves 4~6

25 分鐘
25 minutes

### 醃料 | Marinade

薑汁 1 湯匙
紹興酒 1 茶匙
生粉 1 茶匙
油 1 茶匙
鹽 1/2 茶匙
糖 1/2 茶匙

1 tbsp ginger juice
1 tsp Shaoxing wine
1 tsp cornstarch
1 tsp oil
1/2 tsp salt
1/2 tsp sugar

### 調味料 | Seasonings

鹽 1/2 茶匙

1/2 tsp salt

### 入廚貼士 | Cooking Tips

- 豬肝帶有血腥味道，所以必須加入薑汁酒才能去除那些異味。
- Ginger wine must be added to remove the bloody smell of the pig's liver.

### 做法 | Method

1. 枸杞洗淨，摘葉，留梗備用。
2. 豬肝洗淨切片，加入醃料拌勻。
3. 枸杞梗、薑片和清水同置煲中煮 10~15 分鐘，取出枸杞梗。
4. 加入半份豬肝煮 10 分鐘，取出棄去。
5. 放入枸杞葉和豬肝煮滾，熄火，加入雞蛋便成。

1. Rinse matrimony vine thoroughly, remove leaves and save the stem for later use.
2. Rinse pig's liver thoroughly, slice, add marinade and stir well.
3. Add matrimony vine stem, ginger slices and water into a pot, boil for 10~15 minutes, discard matrimony vine stem.
4. Add half portion of the pig's liver, boil for 10 minutes, then discard.
5. Add matrimony vine leaves and pig's liver, bring to a boil. Remove from heat. Stir in egg and stir well, then ready to serve.

3~4 人
Serves 3~4

20~30 分鐘
20~30 minutes

# 益母草豬肝肉片湯

## Pig's Liver and Pork Slice Soup with Motherwort Herb

### 材料 | Ingredients

益母草（洗淨）450 克
豬肝 150 克
瘦豬肉 150 克
蜜棗 2~3 粒
清水 2 公升

450g motherwort herb (rinsed thoroughly)
150g pig's liver
150g lean pork
2~3 candied dates
2 litres water

### 醃料（豬肝）| Marinade for Pig's Liver

| | |
|---|---|
| 薑汁 1 湯匙 | 1 tbsp ginger juice |
| 紹興酒 1 茶匙 | 1 tsp Shaoxing wine |
| 生粉 1 茶匙 | 1 tsp cornstarch |
| 油 1 茶匙 | 1 tsp oil |
| 鹽 1/2 茶匙 | 1/2 tsp salt |
| 糖 1/2 茶匙 | 1/2 tsp sugar |

### 醃料（瘦肉）| Marinade for Lean Pork

| | |
|---|---|
| 薑汁 1 湯匙 | 1 tbsp ginger juice |
| 紹興酒 1 茶匙 | 1 tsp Shaoxing wine |
| 生粉 1 茶匙 | 1 tsp cornstarch |
| 油 1 茶匙 | 1 tsp oil |
| 鹽 1/2 茶匙 | 1/2 tsp salt |
| 糖 1/2 茶匙 | 1/2 tsp sugar |

### 做法 | Method

1. 豬肝洗淨，切片，加入醃料拌勻。
2. 瘦肉洗淨，切片，加入醃料拌勻。
3. 蜜棗和清水同煮 5 分鐘，加入豬肝和瘦肉煮 5~10 分鐘。
4. 加入益母草煮3~5分鐘即可。

#### 入廚貼士 | Cooking Tips

- 這款湯適合女性飲用，不過益母草帶點苦味，所以加入蜜棗中和味道。
- This soup is suitable for women, candied dates are used to neutralize the bitter taste of motherwort herb.

1. Rinse pig's liver thoroughly, slice and marinate.
2. Rinse lean pork, slice and add marinade.
3. Boil candied dates and water together for 5 minutes, add pig's liver and lean pork and boil for 5~10 minutes.
4. Add motherwort herb and boil for 3~5 minutes, serve.

# 雜錦瓜粒湯

## Assorted Soup with Melon Dice

### 材料 | Ingredients

冬瓜 300 克
瘦豬肉（切丁粒）200 克
雞肉（切丁粒）150 克
鮮蓮子 50 克
鮮冬菇 / 草菇（汆水）50 克
蝦肉（切粒）50 克
蟹肉（蒸熟）50 克
青豆或香花菜 20 克
瑤柱（已浸泡）10 克
鮮百合（洗淨）1 粒
清水 2 公升

300g winter melon
200g lean pork (diced)
150g chicken meat (diced)
50g fresh lotus seed
50g fresh black mushrooms/ straw mushrooms (scalded)
50g shrimps (diced)
50g crab meat (steamed)
20g green pea or mint leaves
10g dried scallop (soaked)
1 fresh lily blub (rinsed thoroughly)
2 litres water

4~6 人
Serves 4~6

25~30 分鐘
25~30 minutes

### 醃料（瘦豬肉）| Marinade for Lean Pork

薑汁 1 湯匙，紹興酒 1 茶匙
生粉 1 茶匙，油 1 茶匙
鹽 1/2 茶匙，糖 1/2 茶匙

1 tbsp ginger juice
1 tsp Shaoxing wine
1 tsp cornstarch, 1 tsp oil
1/2 tsp salt, 1/2 tsp sugar

### 醃料（雞肉）| Marinade for Chicken Meat

雞蛋白 1 隻，生粉 1 茶匙
油 1 茶匙，鹽 1/2 茶匙
糖 1/2 茶匙

1 egg white , 1 tsp cornstarch
1 tsp oil,1/2 tsp salt
1/2 tsp sugar

### 做法 | Method

1. 冬瓜去皮，切丁粒。
2. 瘦豬肉和雞肉分別加入醃料醃片刻。
3. 清水及已浸泡瑤柱連水同置煲中煮滾，加入雞肉和瘦豬肉煮 15 分鐘。
4. 加入冬瓜粒煮 5~10 分鐘，再倒入其餘材料同煮 10 分鐘，即成。

#### 入廚貼士 | Cooking Tips

- 不用冬瓜，可改用合掌瓜或魚翅瓜，味道同樣美味。
- Winter melons may be replaced by chayote (closed palms squash) or spaghetti squash, it will be equally delicious.

1. Peel and dice winter melon.
2. Add marinade to lean pork and chicken meat separately.
3. Add water and soaked dried scallops with water into a pot, bring to a boil. Add chicken meat and lean pork, boil for 15 minutes.
4. Add winter melon dice and boil for 5~10 minutes. Add remaining ingredients and boil for 10 minutes, serve.

4~6 人
Serves 4~6

30~35 分鐘
30~35 minutes

# 魚翅瓜肉片湯

# Spaghetti Squash and Pork Slice Soup

## 材料 | Ingredients

魚翅瓜 1 個
粟米 1 條
瘦豬肉 200 克
大頭菜 5~10 克
薑片 1~2 片
清水 1 1/2 公升

1 spaghetti squash
1 stalk corn
200g lean pork
5~10g preserved turnip
1~2 slices giner
1 1/2 litres water

### 醃料 | Marinade

薑汁 1 湯匙
紹興酒 1 茶匙
生粉 1 茶匙
油 1 茶匙
鹽 1/2 茶匙
糖 1/2 茶匙

1 tbsp ginger juice
1 tsp Shaoxing wine
1 tsp cornstarch
1 tsp oil
1/2 tsp salt
1/2 tsp sugar

**入廚貼士 | Cooking Tips**

- 魚翅瓜的瓜皮很硬，不易削去，所以也可以連皮一起煲湯。
- The skin of the spaghetti squash is rather hard and not easy to be removed, hence, you may boil the soup with the skin.

### 做法 | Method

1. 瘦豬肉洗淨，切片，加入醃料拌勻。
2. 魚翅瓜去皮，切塊。大頭菜用清水浸 10 分鐘，再用鹽擦洗，沖淨。
3. 魚翅瓜、大頭菜、薑片和清水同置煲中煮 20~25 分鐘。
4. 加入粟米和瘦豬肉同煮 8~10 分鐘，即成。

1. Rinser lean pork thoroughly, slice, add marinade and stir well.
2. Peel spaghetti squash and cut into pieces. Soak preserved turnip in water for 10 minutes, then rub and rinse with salt, rinse with water thoroughly.
3. Put spaghetti squash, preserved turnip, ginger slices and water into a pot and boil for 20~25 minutes.
4. Add corn and lean pork and boil together for 8~10 minutes, serve.

# 大葱牛肉豆腐湯

## Beef and Beancurd Soup with Peking Scallion

### 材料 | Ingredients

牛肉 150 克
北京大葱 1~2 條
豆腐（切件）1 磚
胡椒粒 1 湯匙
蒜蓉 1 茶匙
清雞湯 500 毫升
清水 500 毫升

150g beef
1~2 stalks Peking scallion
1 cube beancurd (cut into pieces)
1 tbsp pepper grains
1 tsp minced garlic
500 ml chicken broth
500 ml water

4~6 人
Serves 4~6

10 分鐘
10 minutes

## 醃料 | Marinade

薑汁 1 湯匙
醬油 1 茶匙
紹興酒 1 茶匙
生粉 1 茶匙
油 1 茶匙
糖 1/2 茶匙

1 tbsp ginger juice
1 tsp soy sauce
1 tsp Shaoxing wine
1 tsp cornstarch
1 tsp oil
1/2 tsp sugar

## 做法 | Method

1. 牛肉洗淨，切片，加入醃料醃 5 分鐘。
2. 大葱洗淨切片，備用。
3. 熱鑊下油 1 湯匙，加入豆腐煎香，盛起。
4. 原鑊下點油，加入蒜蓉和胡椒粒爆香，加入大葱，下清雞湯和清水同煮滾。
5. 加入牛肉和豆腐，待滾起，即成。

1. Rinse beef thoroughly, slice, add marinade and leave for 5 minutes.
2. Rinse Peking scallion thoroughly, section.
3. Add 1 tbsp of oil to a heated wok, add beancurd and shallow-fry until fragrant, transfer to a dish.
4. Add some oil to the same wok, add minced garlic and pepper grains and stir-fry until fragrant. Add Peking scallion, chicken broth and water, bring to a boil.
5. Add beef and beancurd and bring to a boil, serve.

4~6 人
Serves 4~6

25~30 分鐘
25~30 minutes

# 雞酒 Chicken Wine

## 材料 | Ingredients

| | |
|---|---|
| 鮮雞（斬件）600 克 | 600g fresh chicken (chopped into pieces) |
| 粉腸 150 克 | 150g pig's small intestine and duodenum |
| 雞雜 2 副 | 2 sets chicken giblets |
| 瘦豬肉（切片）150 克 | 150g lean pork (sliced) |
| 豬肝（切片）150 克 | 150g pg's liver (sliced) |
| 薑絲 40 克 | 40g shredded ginger |
| 木耳絲 20 克 | 20g shredded wood fungus |
| 糯米酒 100 毫升 | 100 ml glutinous rice wine |
| 清水 2 公升 | 2 litres water |

## 醃料（鮮雞、雞雜、瘦肉和豬肝）| Marinade (For Fresh Chicken, Chicken Giblet, Lean Pork and Pig's Liver)

| | |
|---|---|
| 薑汁 1 湯匙 | 1 tbsp ginger juice |
| 紹興酒 1 茶匙 | 1 tsp Shaoxing wine |
| 生粉 1 茶匙 | 1 tsp cornstarch |
| 油 1 茶匙 | 1 tsp oil |
| 鹽 1/2 茶匙 | 1/2 tsp salt |
| 糖 1/2 茶匙 | 1/2 tsp sugar |

## 做法 | Method

1. 粉腸用薑粒或蒜粒通洗，去脂肪，切段。
2. 雞雜用生粉和油擦洗，沖淨，加入調味。
3. 把所有材料分別用油爆炒至半熟。
4. 熱鑊下油，加入薑絲炒透，再把所有材料（除了糯米酒）回鑊，灒酒，注入清水煮 20~30 分鐘。加入糯米酒煮滾，即可。

1. Rinse pig's small intestine and duodenum with ginger or garlic dice by going through them. Remove the fat and cut into sections.
2. Wipe and rinse chicken giblet with cornstarch and oil, flush with water thoroughly, add marinade.
3. Heat a wok and stir-fry all ingredients with oil separately to half done.
4. Add oil to a heated wok, add shredded ginger and stir-fry thoroughly, then put all ingredients (except the glutinous rice wine) back to the wok. Sprinkle with wine, add water and boil for 20~30 minutes. Add glutinous rice wine, bring to a boil and serve.

# 雞球大蝦時菜湯

## Chicken Ball and King Prawn Soup with Vegetables

### 材料 | Ingredients

雞翼 450 克
大蝦 300 克
時菜 300 克
清雞湯 500 毫升
清水 500 毫升

450g chicken wings
300g king prawns
300g seasonal vegetables
500 ml chicken broth
500 ml water

4~6 人
Serves 4~6

15~20 分鐘
15~20 minutes

## 醃料 | Marinade

薑汁 1 湯匙
紹興酒 1 茶匙
生粉 1 茶匙
油 1 茶匙
鹽 1/2 茶匙
糖 1/2 茶匙

1 tbsp ginger juice
1 tsp Shaoxing wine
1 tsp cornstarch
1 tsp oil
1/2 tsp salt
1/2 tsp sugar

### 入廚貼士 | Cooking Tips

- 不用雞翼的話，可改用雞胸，不過雞胸肉會不夠幼滑。
- Chicken brisket may be used to replace chicken wing, but the meat will not be smooth enough.

## 做法 | Method

1. 雞翼去骨，洗淨抹乾，加醃料拌勻。
2. 大蝦去頭、去殼、去腸，備用。
3. 清雞湯和清水煮滾，加入雞球和時菜煮 8~10 分鐘。
4. 再加入大蝦煮 3~4 分鐘，即可。

1. Remove bones from chicken wings. Rinse and wipe dry, add marinade and stir well.
2. Remove prawns' heads, shells and intestines, set aside.
3. Add chicken broth and water into a pot, bring to a boil. Add chicken balls and vegetables, boil for 8~10 minutes.
4. Add prawns and boil for 3~4 minutes, serve.

4~6 人
Serves 4~6

30~40 分鐘
30~40 minutes

# 菜肉餃子清雞湯

# Chicken Broth with Vegetable and Meat Dumplings

## 材料 | Ingredients

| | |
|---|---|
| 鮮雞 1/2 隻 | 1/2 fresh chicken |
| 鹹豬肉 80 克 | 80g salted pork |
| 粟米 1 條 | 1 stalk corn |
| 甘筍 1 條 | 1 carrot |
| 時菜 300 克 | 300g seasonal vegetables |
| 菜肉餃子 12 隻 | 12 vegetable and meat dumplings |
| 清水 2 公升 | 2 litres water |

## 做法 | Method

1. 鮮雞洗淨，斬件，汆水，過冷。
2. 鹹豬肉汆水，過冷，切片。
3. 鮮雞、鹹豬肉、粟米、甘筍和清水同置煲中煮 30 分鐘。
4. 加入時菜和菜肉餃子煮 8~10 分鐘，即成。

1. Rinse chicken thoroughly and chop into pieces, scald and rinse with cold water.
2. Scald salted pork, rinse with cold water, slice.
3. Put chicken, salted pork, corn, carrot and water into a pot, boil for 30 minutes.
4. Add vegetables and vegetable and meat dumplings, boil for 8~10 minutes and serve.

# 千張鹹肉津白湯

# Thin Dried Beancurd, Salted Pork and Tianjin Cabbage Soup

4~6 人
Serves 4~6

30~40 分鐘
30~40 minutes

## 材料 | Ingredients

| | |
|---|---|
| 鹹豬肉 80 克 | 80g salted pork |
| 百頁 50 克 | 50g thin dried beancurd |
| 鮮枝竹 4~6 條 | 4~6 fresh beancurd sticks |
| 津白（切段）300 克 | 300g Tianjin cabbage (sectioned) |
| 鮮菇（汆水）80 克 | 80g fresh mushroom (scalded) |
| 甘筍（切片）40 克 | 40g carrot (sliced) |
| 薑 2~3 片 | 2~3 slices ginger |
| 清雞湯 500 毫升 | 500 ml chicken broth |
| 清水 500 毫升 | 500 ml water |

## 做法 | Method

1. 百頁用梳打水浸片刻，沖洗乾淨，汆水，過冷，打結。
2. 鹹豬肉汆水，過冷，切片。
3. 清雞湯、清水、薑和甘筍煮滾，加入鹹肉片煮 20 分鐘。
4. 加入津白、鮮菇、鮮枝竹和百頁同煮 10~15 分鐘，即成。

1. Soak thin dried beancurd in soda water for a while. Rinse thoroughly, scald, rinse with cold water. Make knots.
2. Scald salted pork, rinse with cold water and slice.
3. Bring chicken broth, water, ginger and carrot to a boil, add salted pork slices and boil for 20 minutes.
4. Add Tianjin cabbage, fresh mushroom, fresh beancurd sticks and thin dried beancurd, boil together for 10~15 minutes, serve.

4~6 人
Serves 4~6

25~30 分鐘
25~30 minutes

# 洋蔥牛肉清湯
# Beef Broth with Onion

### 材料 | Ingredients

免治牛肉 600 克
洋葱 2 個
雞蛋 2 隻
蒜蓉 1 茶匙
清水 1 1/2 公升

600g minced beef
2 onions
2 eggs
1 tsp minced garlic
1 1/2 litres water

### 醃料 | Marinade

鹽 1 茶匙
胡椒粉適量

1 tsp salt
pinch of pepper

### 做法 | Method

1. 免治牛肉加入醃料拌勻。加入雞蛋拌勻，分成 3 個大肉丸，蛋殼留用。
2. 洋葱去衣，切絲。熱鑊下油，加入蒜蓉和洋葱炒香，注入清水煮滾，熄火停 5 分鐘。
3. 轉慢火，牛肉丸和蛋殼放入洋葱水中煮 20~30 分鐘。

1. Add marinade to minced beef, stir well. Add eggs and stir well, divide into 3 big meat balls. Keep egg shells for later use.
2. Peel onion and shred. Add oil to a heated wok, add minced garlic and onion, stir-fry till fragrant. Add water and bring to a boil, remove from heat and leave for 5 minutes.
3. Turn to low heat, add beef meat balls and egg shells into the onion soup, boil for 20~30 minutes.

# 蘿蔔牛腩碎清湯

# Minced Beef Brisket Broth with Radish

4~6 人
Serves 4~6

1 小時
1 hour

## 材料 | Ingredients

白蘿蔔 1 條
牛腩碎 600 克
蒜頭 1~2 粒
薑 2~3 片
八角 1~2 粒
香葉 1 片
日本清酒 50 毫升
清水 3 公升

1 radish
600g beef brisket
1~2 cloves garlic
2~3 slices ginger
1~2 pc star anise
1 bay leaf
50 ml Japanese sake
3 litres water

### 入廚貼士 | Cooking Tips

- 牛腩碎可用磨豉醬加蠔油芡汁燴煮，便可當一味家常小菜。
- You may stew the minced beef brisket with fermented bean paste and oyster sauce, it will then be a delicious dish.

## 做法 | Method

1. 白蘿蔔去皮，切塊。
2. 牛腩碎洗淨，去肥油，汆水 10~15 分鐘，過冷。
3. 熱鑊下油，爆香蒜頭、薑和八角，注入清水、香葉和放入牛腩碎以大火煮滾。轉中火煮 50 分鐘或至牛腩腍軟。
4. 加入蘿蔔煮 10 分鐘或透明，熄火後加入清酒即成。

1. Peel radish and chop into pieces.
2. Rinse minced beef brisket thoroughly and remove the fat. Scald for 10~15 minutes and rinse with cold water.
3. Add oil to a heated wok, stir-fry the garlic, ginger and star anise till fragrant. Add water, bay leave and minced beef brisket, cook over high heat and bring to a boil. Turn to medium heat and bring to a boil, cook for 50 minutes or until the beef brisket turns soft.
4. Add radish and boil for 10 minutes or until transparent. Remove from heat, add sake and serve.

4~6 人
Serves 4~6

25~30 分鐘
25~30 minutes

# 魚丸牛肉湯 Fish Ball and Beef Soup

### 材料 | Ingredients

牛肉 300 克
鯇魚 / 鯪魚 1 條
洋葱 1 個
時菜 300 克
蒜蓉 1 茶匙
清水 1 公升

300g beef
1 mackeral/dace
1 onion
300g seasonal vegetables
1 tsp minced garlic
1 litre water

### 醃料 | Marinade

雞蛋白 1 隻
糖 1 茶匙
生粉 1 茶匙
胡椒粉 1 茶匙
鹽 1/2 茶匙
麻油少許
清水 1 湯匙

1 egg white
1 tsp sugar
1 tsp cornstarch
1 tsp pepper
1/2 tsp salt
some sesame oil
1 tbsp water

### 調味料 | Seasonings

鹽 1/2 茶匙
胡椒粉適量

1/2 tsp salt
pinch of pepper

#### 入廚貼士 | Cooking Tips

- 牛肉浸水會有點血腥味道，可加點酒來去除腥味。
- Some wine can be added to remove the bloody smell of beef after soaking in water.

### 做法 | Method

1. 牛肉洗淨，切片，加入清水浸 30 分鐘，隔渣。
2. 鯪魚用湯匙把魚肉刮出魚肉，加入醃料拌勻，攪成有彈性和黏度。
3. 熱鑊下油，加入蒜蓉和洋葱炒香，注入牛肉水煮滾，放入魚丸和時菜煮 5~10 分鐘。

1. Rinse beef thoroughly and slice, add water and soak for 30 minutes, strain off the residue.
2. Remove the flesh from the body of the dace with a spoon. Add marinade to the dace, mix well and stir until springy and sticky.
3. Add oil to a heated wok, add minced garlic and onion and stir-fry until fragrant. Add beef soup and bring to a boil. Then add fish balls and vegetables and boil for 5~10 minutes.

# 香芹肉碎蠔仔湯

# Oyster Soup with Minced Meat and Celery

## 材料 | Ingredients

| | |
|---|---|
| 蠔仔 200 克 | 200g small oysters |
| 免治豬肉 80 克 | 80g minced pork |
| 中國芹菜（切粒）1 棵 | 1 stalk Chinese celery (diced) |
| 芫荽（切碎）1~2 棵 | 1~2 stalks coriander (minced) |
| 冬菜 1 湯匙 | 1 tbsp preserved Chinese cabbage |
| 蒜蓉 1 茶匙 | 1 tsp minced garlic |
| 胡椒粒 1 茶匙 | 1 tsp pepper grain |
| 清雞湯 500 毫升 | 500 ml chicken broth |
| 清水 500 毫升 | 500 ml water |

4~6 人
Serves 4~6

15 分鐘
15 minutes

### 醃料 | Marinade

紹興酒 1 茶匙，油 1/2 茶匙
鹽 1/4 茶匙，糖 1/4 茶匙
醬油 1/4 茶匙，生粉 1/4 茶匙

1 tsp Shaoxing wine, 1/2 tsp oil
1/4 tsp salt, 1/4 tsp sugar
1/4 tsp soy sauce
1/4 tsp cornstarch

### 入廚貼士 | Cooking Tips

- 蠔仔必須小心清洗，特別是蠔裙邊的位置是細菌溫床，加上蠔堆中偶然會藏有碎殼和沙石。
- Wash small oysters carefully, especially the edge position where is the incubator for bacteria. Besides, there may be broken shells and sand in the oysters.

### 調味料 | Seasonings

| | |
|---|---|
| 金蒜片 1~2 湯匙 | 1~2 tbsps golden garlic slices |
| 鹽 1/2 茶匙 | 1/2 tsp salt |
| 胡椒粉適量 | pinch of pepper |

### 做法 | Method

1. 蠔仔用生粉和油擦洗乾淨，沖水，瀝乾。
2. 免治豬肉加入醃料拌勻，備用。
3. 熱鑊下油，放入蒜蓉、芹菜梗和芫荽頭炒香，倒入免治豬肉，灒酒，注入清雞湯和清水煮滾，取去芫荽頭。
4. 放入蠔仔煮5分鐘，下冬菜和調味拌勻，吃時加入芫荽碎和金蒜片。

1. Rub and rinse the oysters with cornstarch and oil thoroughly, rinse with water and drain.
2. Mix minced pork with marinade and stir well.
3. Add oil to a heated wok, add minced garlic, celery stalk and coriander stalks, stir-fry until fragrant. Add minced pork, sprinkle with wine, pour in chicken broth and water, bring to a boil and remove coriander stalks.
4. Add small oysters and cook for 5 minutes. Add preserved Chinese cabbage and seasonings, stir well, serve with minced coriander and golden garlic slices.

4~6 人
Serves 4~6

30~40 分鐘
30~40 minutes

# 鹹菜排骨涼瓜湯

# Sparerib Soup with Chinese Pickled Vegetable and Bitter Melon

## 材料 | Ingredients

| | |
|---|---|
| 排骨 200 克 | 200g spareribs |
| 鹹菜 150 克 | 150g salted vegetable |
| 黃豆（已浸泡）40 克 | 40g soy beans (soaked) |
| 涼瓜 1 個 | 1 bitter melon |
| 蒜頭 1~2 粒 | 1~2 cloves garlic |
| 清水 1 1/2 公升 | 1 1/2 litres water |

## 做法 | Method

1. 涼瓜去籽，切片。鹹菜用清水浸 1 小時，切片。
2. 熱鑊下油，加入蒜頭爆香，加入排骨略煎香。
3. 注入清水，放入鹹菜和黃豆煮 30 分鐘。加入涼瓜煮 10 分鐘，即成。

1. Remove seeds from bitter melon and slice. Soak salted vegetables in water for 1 hour, drain and slice.
2. Add oil to a heated wok, add garlic and stir-fry till fragrant. Add spareribs and shallow-fry until golden.
3. Add water, salted vegetable and soy beans and boil for 30 minutes. Add bitter melon and boil for 10 minutes, serve.

# 芥菜豬雜湯

# Mustard Greens and Pig's Offal Soup

4~6 人
Serves 4~6

1 小時
1 hour

### 材料 | Ingredients

芥菜 300 克
豬肚（急凍）1 個
豬腸（急凍）1 條
瘦豬肉 150 克
鹹蛋 1~2 隻
薑 2~3 片
清水 2 公升

300g mustard greens
1 pig's stomach (frozen)
1 pig's intestine (frozen)
150g lean pork
1~2 salted eggs
2~3 slices ginger
2 litres water

### 醃料 | Marinade

薑汁 1 湯匙
紹興酒 1 茶匙
生粉 1 茶匙
油 1 茶匙
鹽 1/2 茶匙
糖 1/2 茶匙

1 tbsp ginger juice
1 tsp Shaoxing wine
1 tsp cornstarch
1 tsp oil
1/2 tsp salt
1/2 tsp sugar

### 做法 | Method

1. 芥菜洗淨，切段。
2. 瘦豬肉洗淨，切片，加醃料拌勻，備用。
3. 豬肚和豬腸汆水，過冷，加清水，大火煮 40 分鐘。
4. 加入肉片煮 5 分鐘，再加入芥菜和鹹蛋煮 5~10 分鐘。

1. Rinse mustard greens thoroughly and cut into sections.
2. Rinse lean pork and slice, add marinade and stir well.
3. Scald pig's stomach and intestine, rinse with cold water. Add water and boil over high heat for 40 minutes.
4. Add pork slices and boil for 5 minutes. Then add mustard greens and salted egg and boil for 5~10 minutes.

**編著**
何美好

**編輯**
紫彤

**美術設計**
YU Cheung

**翻譯**
黃卿

**攝影**
幸浩生

**出版者**
萬里機構出版有限公司
香港鰂魚涌英皇道1065號東達中心1305室
電話：2564 7511
傳真：2565 5539
電郵：info@wanlibk.com
網址：http://www.wanlibk.com
http://www.facebook.com/wanlibk

**發行者**
香港聯合書刊物流有限公司
香港新界大埔汀麗路36號
中華商務印刷大廈3字樓
電話：2150 2100
傳真：2407 3062
電郵：info@suplogistics.com.hk

**承印者**
美雅印刷製本有限公司

**出版日期**
二零一八年五月第一次印刷

萬里機構

萬里 Facebook


ISBN 978-962-14-6711-9